discovermore

Forms of Energy

What Is Electrical Energy?

Virginia Elliott

Britannica Educational Publishing

IN ASSOCIATION WITH

THE ROSEN PUBLISHING GROUP

Published in 2026 by Britannica Educational Publishing (a trademark of Encyclopædia Britannica, Inc.) in association with The Rosen Publishing Group, Inc.
2544 Clinton Street, Buffalo, NY 14224

Distributed exclusively by Rosen Publishing.
To see additional Britannica Educational Publishing titles, go to rosenpublishing.com.

Editor: Danielle Haynes
Book Design: Michael Flynn

Photo Credits: Cover pumnaruk/Shutterstock.com; (series background) Dai Yim/Shutterstock.com; p. 4 MMPhoto21/Shutterstock.com; p. 5 YesPhotographers/Shutterstock.com; p. 7 (top) Inkoly/Shutterstock.com; p. 7 (bottom) digitaliensp b/Shutterstock.com; p. 8 kryzhov/Shutterstock.com; p. 9 Catatan Bunda/Shutterstock.com; p. 10 Flegere/Shutterstock.com; p. 11 ungvar/Shutterstock.com; p. 12 Boyloso/Shutterstock.com; p. 13 Erica Finstad/Shutterstock.com; p. 14 Vasin Lee/Shutterstock.com; p. 15 Designua/Shutterstock.com; p. 16 Lopolo/Shutterstock.com; p. 17 Stella_E/Shutterstock.com; p. 18 Song_about_summer/Shutterstock.com; p. 19 Gorodenkoff/Shutterstock.com; p. 20 (left) IM Imagery/Shutterstock.com; p. 20 (right) Pixel-Shot/Shutterstock.com; p. 21 Jabeen Fatima M J/Shutterstock.com; p. 22 GheorgheGarcu/Shutterstock.com; p. 23 2p2play/Shutterstock.com; p. 24 LouieLea/Shutterstock.com; p. 25 Zaporizhzhia vector/Shutterstock.com; p. 26 Pi-Lens/Shutterstock.com; p. 27 chrisdorney/Shutterstock.com; p. 28 https://en.wikipedia.org/wiki/File:Joseph_Siffrein_Duplessis_-_Benjamin_Franklin_-_Google_Art_Project.jpg; p. 29 Everett Collection/Shutterstock.com.

Cataloging-in-Publication Data

Names: Elliott, Virginia (Children's writer).
Title: Electrical energy / Virginia Elliott.
Description: New York : Britannica Educational Publishing, in association with Rosen Educational Services, 2026. | Series: Discover more: forms of energy | Includes glossary and index.
Identifiers: ISBN 9781641905633 (library bound) | ISBN 9781641905626 (pbk) | ISBN 9781641905640 (ebook)
Subjects: LCSH: Electricity--Juvenile literature. | Electric power--Juvenile literature.
Classification: LCC QC753.7 E45 2026 | DDC 537--dc23

Manufactured in China

CPSIA Compliance Information: Batch #QWBRIT26. For further information contact Rosen Publishing at 1-800-237-9932.

Contents

Energy Basics

Energy is the ability to do work. It's another word for power. Without energy, machines would not be able to work. Plants and animals would not be able to live and grow. There are many different forms of energy. Heat and light are examples of energy you can find in the world.

Energy is typically measured in watts and less commonly in joules.

A bolt of lightning is about 300 million volts of power, while an electrical outlet in your home typically can handle about 120 volts.

Another form of energy that most people are familiar with is electrical energy, or electricity. It is very important to people. Without electrical energy, we would not have many conveniences. From lightbulbs to televisions, electrical energy powers all sorts of things.

Electricity can be formed using machines, but it is also found in nature. A bolt of lightning is electrical energy that is suddenly released in the atmosphere. It has many times more power than any of the appliances or machines that run in your house.

Consider This

Before lightbulbs were invented, people used fire to see in the dark. What forms of energy are released from fire?

Atoms and Energy

Extremely small particles called electrons are important to the formation of electrical energy. They are associated with the tiny building blocks of matter called atoms. Everything in the universe is made up of atoms.

At the center of an atom is a nucleus. The nucleus consists of particles called protons and neutrons. Protons carry a positive electrical charge, and neutrons carry no electrical charge.

Surrounding the nucleus is a cloud of particles called electrons. Electrons carry a negative electrical charge. They constantly travel in circular paths around the nucleus.

Sometimes electrons from one atom will move to another atom. The moving electrons are electricity. Scientists now understand how to cause electrons to move and how to control where electrons travel. Today, electricity can be formed and then delivered across small and great distances. The faster these electrons move, the more electricity they carry.

Atom Structure

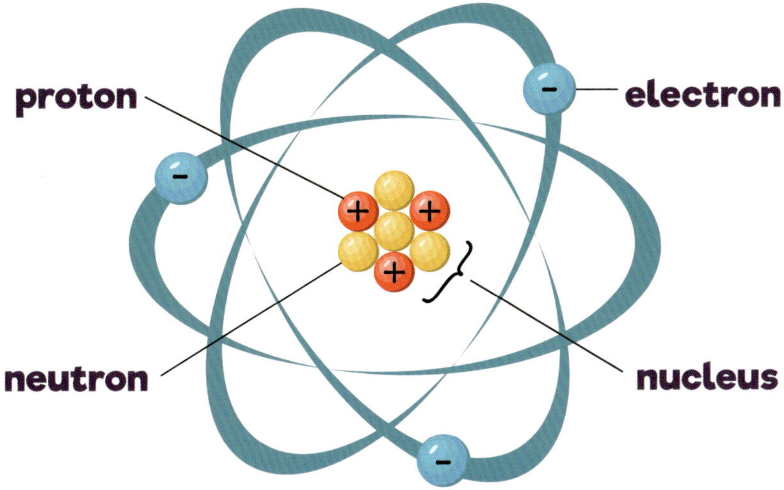

Electrons travel around the nucleus of an atom in paths called electron shells.

compare and contrast

What do protons have in common with electrons and neutrons? In what ways are they different?

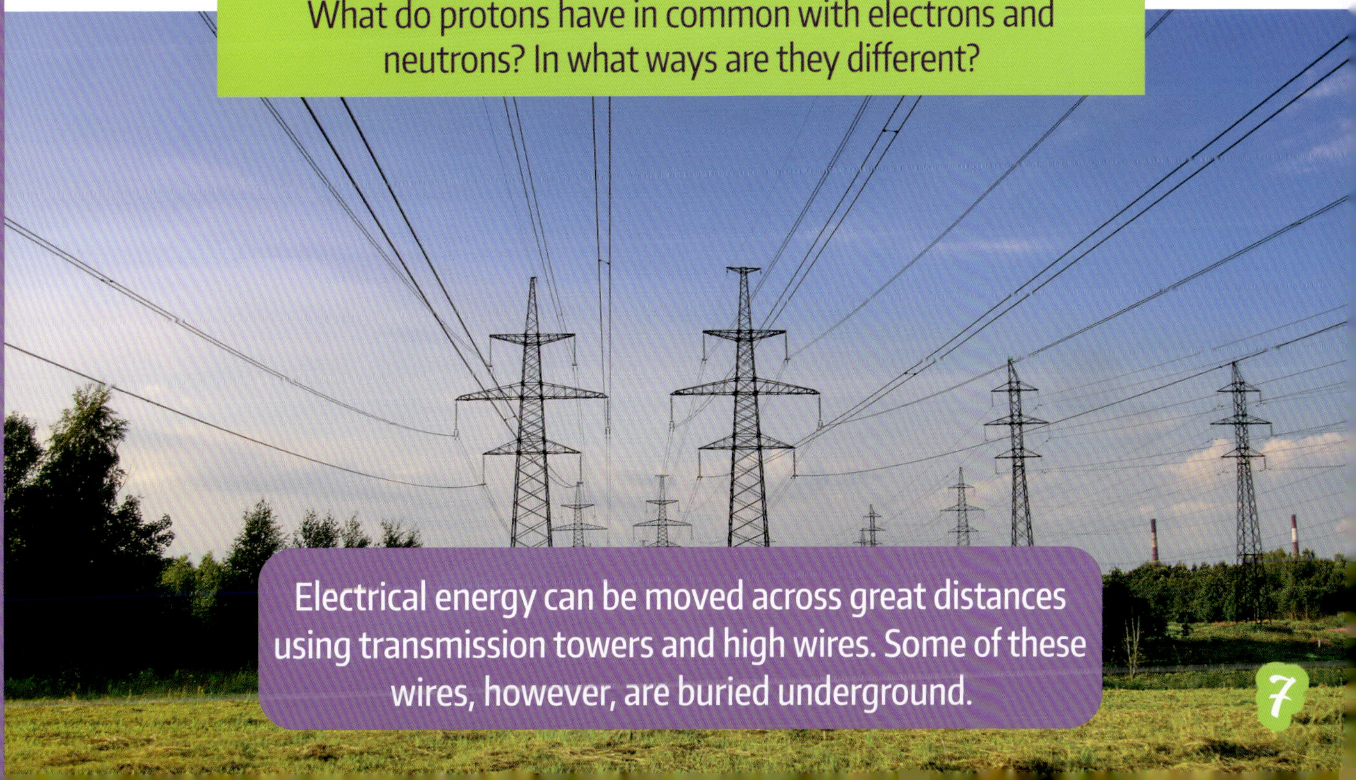

Electrical energy can be moved across great distances using transmission towers and high wires. Some of these wires, however, are buried underground.

Magnetism

The magnetic charges of atomic particles act just like the magnets you use on your refrigerator. Negative and positive charges attract each other. However, two positive charges, or two negative charges, **repel** each other. Electricity results when electrons (which have negative charges) are pushed and pulled from atom to atom.

Some electromagnets are so powerful they can lift huge pieces of metal. These magnets can be turned on and off to lift and drop piles of junk on command.

You can magnetize an iron nail by wrapping copper wire around it and attaching each end of the wire to the ends of a battery.

The push and pull of negative and positive charges causes magnetism. Magnetism is a basic force of nature, like electricity and gravity. Spinning electrons form tiny magnetic forces. Sometimes many of the electrons in an object spin in the same direction. In these cases, all the tiny magnetic forces from the electrons add up to make the object one big magnet. Magnets have many uses. They hold papers on refrigerator doors. They are also used in credit cards, stereo speakers, and compasses.

Electricity can create magnets. As electrons move through a piece of wire, they have the same effect as electrons spinning around the nucleus of an atom. This creates an electromagnet. You could make a simple electromagnet at home!

WORD WISE

TO REPEL MEANS TO PUSH AWAY.

Insulators and Conductors

Electricity moves through some materials better than others. Materials that allow electricity to pass through them easily are called electrical conductors. Many metals, such as iron, steel, copper, and aluminum, are conductors. Equipment that is used to conduct electricity is often made up of metal parts.

Electric circuits are paths for moving electricity. In a simple circuit, copper wire conducts electricity from the energy source (such as a battery) to an electrical device (such as a lamp, motor, or bell).

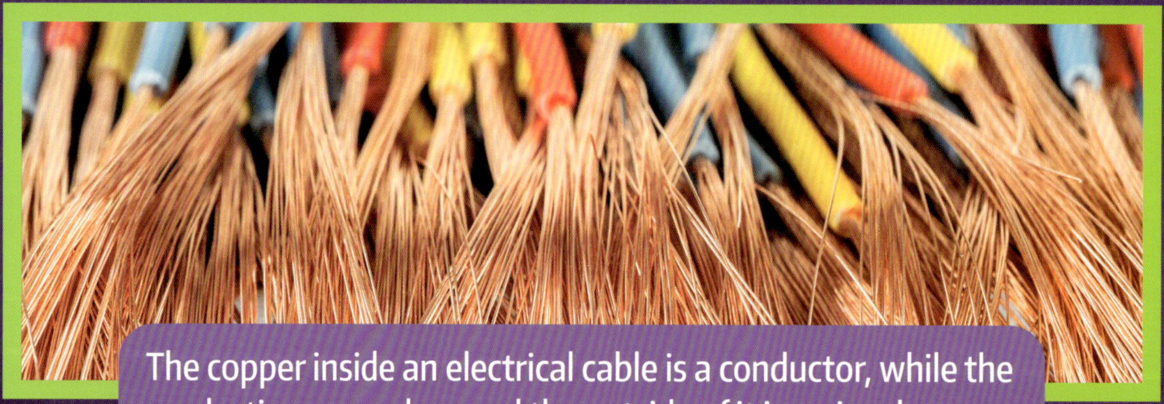

The copper inside an electrical cable is a conductor, while the plastic wrapped around the outside of it is an insulator.

During a storm, it's important to stay away from downed electrical wires. The water on the ground is an electrical conductor.

Materials that do not conduct electricity are called electrical insulators. Examples of insulators include plastic, rubber, wood, and glass. Air is also an insulator.

Most electrical objects are made using insulators to keep them safe. Plugs, for example, have plastic cases. Electrical wires are wrapped in plastic, which is flexible as well as insulating. Electricians often wear rubber-lined gloves to protect themselves.

Consider This

Without insulators, many common devices people use would be dangerous to touch. Can you name three electrical devices you touch every day?

Static

Sometimes, electrons can build up in an object and cause static electricity. Most objects have an equal amount of positive and negative charges, so they are considered neutral. They do not push or pull on each other at all electrically. When there are extra electrons, two such objects can push or pull on each other because they are no longer neutral.

Static electricity can cause pet fur or other items to stick to clothes. People use dryer sheets to fix this problem when washing clothes.

Your hair might stand on end after sliding down a plastic slide or jumping on a trampoline.

Static electricity can cause interesting effects, like sparks or lightning bolts, when it is released. Sometimes the extra electrons build up by rubbing one object against another. For example, when one rubs a balloon against one's hair, electrons move from the balloon to the hair. Because the hairs then all have extra electrons, which all have the same kind of charge, they try to fly away from each other and end up sticking into the air like spikes. You're more likely to have staticky hair in the winter when there's not as much humidity in the air.

Consider This

Static electricity pulls objects together or pushes them apart. What other force acts in a similar way?

Lightning

Electricity happens naturally, not just from human-made sources. Lightning is a natural form of electricity. Water droplets and ice particles inside a cloud carry electrical charges. Some charges are positive, and others are negative. Lightning occurs usually when too many negative charges build up in a cloud. To balance out these negative charges, positive charges form under the cloud on the ground. Because opposite charges attract, the negative charges in the cloud want to join the positive charges below.

Scientists estimate there are nearly 8 million lightning strikes per day across the globe!

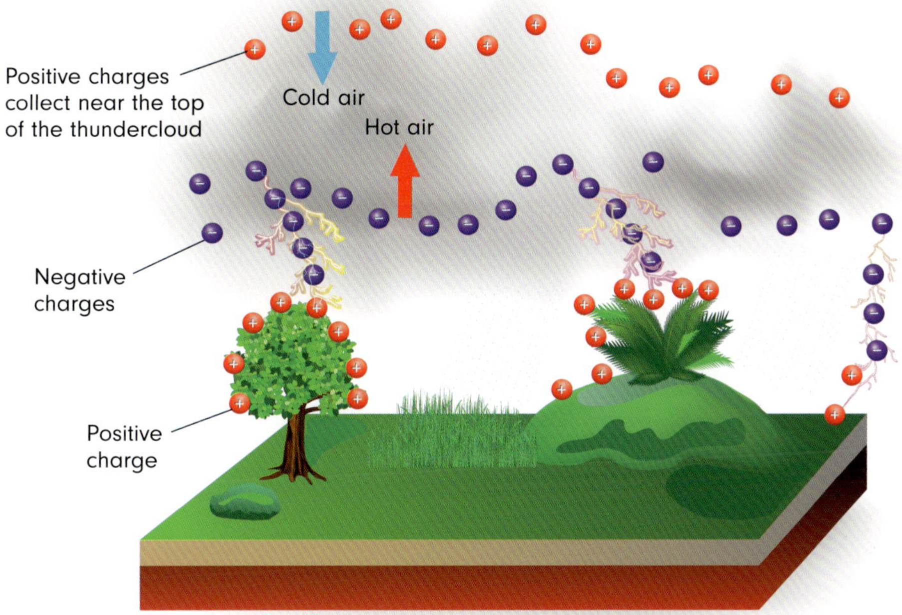

You can never predict exactly where a lightning flash will strike, so it's best to stay indoors during storms.

It's difficult for the charges to unite because air is an insulator. As the cloud grows, however, so does the strength of the charges. Eventually the charges overpower the air, and the cloud releases a strong, negatively charged **electrical current**.

As the negative current heads toward the ground, a positively charged current jumps from the ground to meet it. When the currents join, a bright flash is created that heads back up toward the cloud. This is a lightning flash. These flashes tend to strike tall objects such as buildings and trees.

WORD WISE

AN ELECTRICAL CURRENT IS A STREAM OF MOVING ELECTRONS.

Electrical Energy in Humans

Believe it or not, electricity flows all throughout your body. Without electrical energy and electricity, you would not be able to read this book. Every action you take is made possible by electrical signals that travel through your body.

Like everything around us, our bodies are made up of atoms. And in those atoms, there are protons, neutrons, and electrons. When the charges of these particles are out of balance, electrons move to other atoms to create balance.

Every action you make, from running to playing basketball, requires an electrical signal from your brain to the rest of your body.

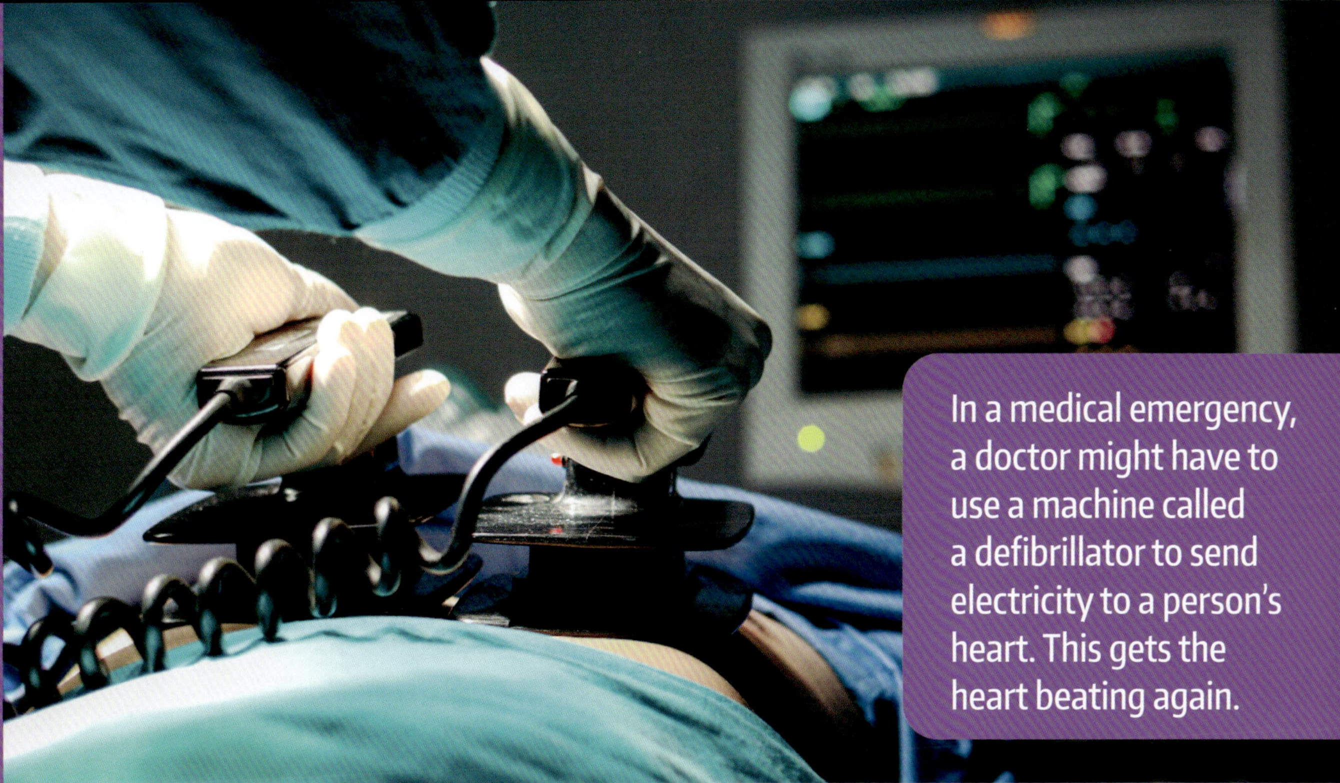

In a medical emergency, a doctor might have to use a machine called a defibrillator to send electricity to a person's heart. This gets the heart beating again.

Electricity travels along nerve cells throughout the body. Electrical signals from your brain tell your body what to do. When you turn the page of a book, your brain is sending electrical signals to your hands. Your eyes send electrical signals back to the brain about what you are reading. The electricity in your body's cells is so weak that it's not dangerous to you or others, unlike lightning.

Consider This

Without electrical energy, we would not be able to think. Is there anything you could do without electrical energy?

Improving Lives

Humans rely on electricity in many ways. We light our homes and schools with lamps that are powered by electricity. The cars and buses we ride in are made up of parts that require electricity.

Learning how to **generate** and deliver electricity has changed the way we live. For example, before the invention of the phone, people had to write letters to communicate over long distances. Now we can call family and loved ones at anytime from anywhere in the world. Like cars, phones use electricity to work.

Before modern airplanes, it took days or weeks to travel to new places.However, electricity has helped to change that. Today, we can travel all over the world in just hours.

Nearly all the equipment in the hospital used to keep patients alive and comfortable uses electricity.

Electricity is also used to save lives. For example, a pacemaker is a device that sends electric signals to a person's heart. The electric signals keep the heart beating regularly if the body cannot do that on its own. Many inventions that have made survival and life easier would not exist without electricity. Think of all the ways life would become more difficult and dangerous if we suddenly lost electricity.

WORD WISE

TO GENERATE SOMETHING IS TO CAUSE IT TO EXIST.

Batteries

Losing power during a storm is annoying—but luckily, we have batteries. Batteries give electric power to flashlights, radios, cell phones, handheld games, and many other types of equipment. A battery is a sort of container that stores energy until it is needed. Chemicals inside the battery store the energy. When the battery is used, the chemical energy changes into electrical energy.

Batteries can be large, such as this one that powers an electric vehicle, or small enough to fit inside a wristwatch.

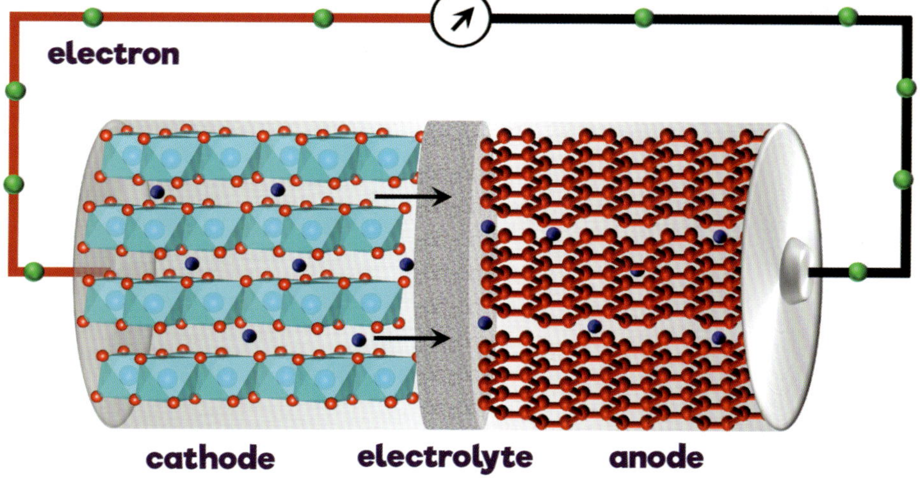

When you put a battery inside an electronic device, it causes a connection between the terminals, and electricity flows.

Inside a battery there are two pieces of metal in a liquid or a paste. The metal parts are called electrodes. The liquid or paste, called an electrolyte, is a mix of chemicals. Each electrode has a point, called a terminal, that sticks out of the battery.

When you turn a flashlight on, an outside wire links the terminals in the battery. Then the chemicals in the electrolyte cause electrons to flow from one electrode to the other. The electric current flowing through the wire makes the flashlight work. Once the chemicals inside the battery are used up, the battery no longer works.

Consider This

Batteries produce different amounts of energy. Why would a car need a larger battery than a flashlight?

Creating Electricity

We use electricity to power many items we use every day. With special machines and **technology**, we are able to produce electricity and use it safely.

A city's power plant uses machines called generators to produce a powerful electric current. Power plants need a source of energy to turn the generators that produce electricity. The moving energy of water or wind is often used to run the generators.

Large wind farms generate electricity by capturing wind energy with wind turbines. Some are on land, while others are built out in the open ocean.

Your phone and laptop have batteries that store the electricity they receive from a power outlet. Older phones and computers could only be used while plugged in to the outlet.

Generators cause a current to flow by moving a magnet past a coil of wire, which pushes electrons through the wires of the coil. The current travels through wires to houses and other buildings. More wires connect to the power outlets in rooms. When a person plugs in an iron or another electric device, the current travels into the device. The current then makes the device work. This same current can also charge the batteries in your phone and laptop.

WORD WISE

TECHNOLOGY IS THE USE OF KNOWLEDGE TO INVENT NEW DEVICES OR TOOLS.

Energy Sources

Most of the world's energy comes from fossil fuels, specifically coal. Coal is a black or brown rock. When a power plant burns coal, it creates heat. The heat then boils water. Steam from the boiling water rotates turbines that power generators.

Hydropower, or waterpower, is another energy source used to power generators. Hydroelectric power plants are usually located in dams that are built across rivers. In a dam, water is collected at a high level and is then led downward through large pipes to a lower level. The falling water rotates turbines that power generators.

Humans have built dams to produce hydroelectric power. This energy comes from the movement of water.

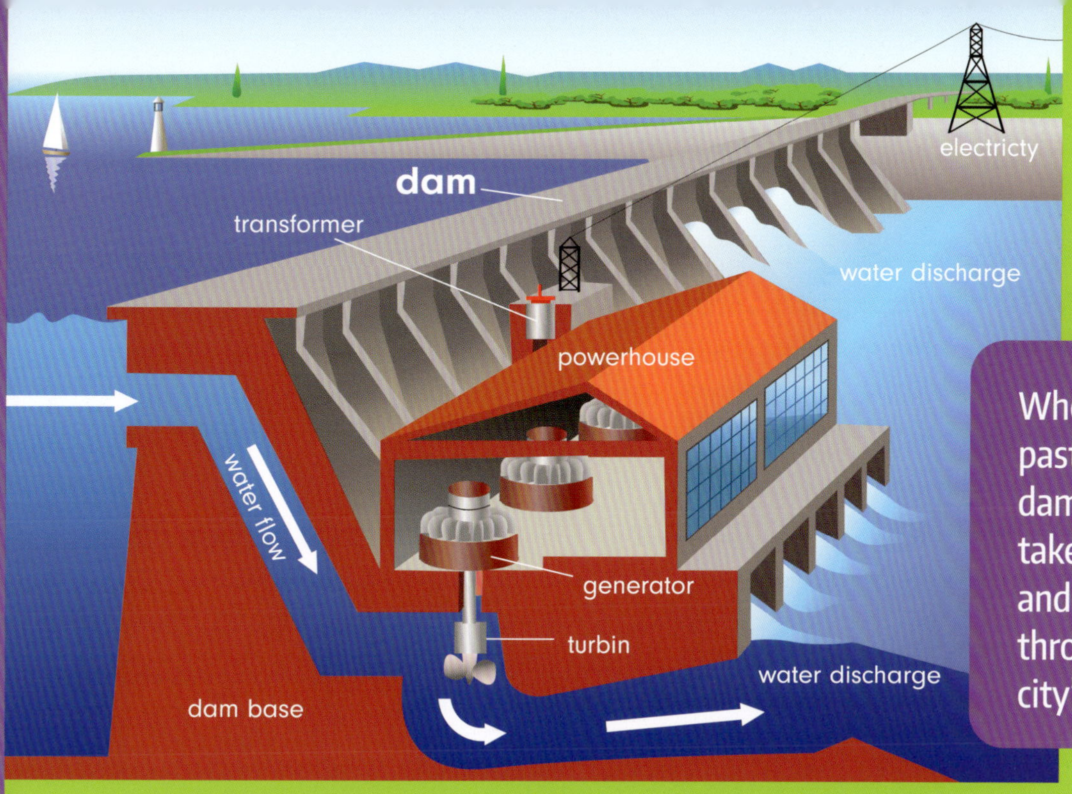

When water moves past a turbine in a dam, the generator takes that energy and sends it through wires to a city's electrical grid.

Other energy sources may also be used to power generators. Coal and other fossil fuels release harmful gases when burned and can be used up. Alternative energy sources, such as water and solar power, cannot be used up and generally do not harm the environment. Over the past several decades, a greater share of the world's energy is coming from renewable sources each year.

Consider This

Windmills can turn moving energy into electricity. What are some benefits of using wind as an energy source?

How Much Electricity?

Did you know it's possible to measure electricity? This ability is important for many reasons. For one, it lets power companies know how much electrical energy a customer uses. Homes and businesses pay their power company for the amount they use. Power companies use electrical meters located outside a home or business to measure the amount of energy that has been used there. Secondly, the measurement of electricity is important for safety reasons. Different devices and products can handle only certain amounts of electricity. Too much electricity can overheat or cause devices to catch on fire or explode.

Many buildings have meters that show the electric company how much power has been used.

Scottish inventor James Watt developed a steam engine that used less fuel than other models.

Electricity is measured in units called watts. The watt is named after the inventor James Watt. A watt is a low amount of power. Small devices, such as lightbulbs, use few watts. Larger appliances and machines require more power. Electricity that larger appliances use is measured in **kilowatt** units. The symbol for watts is W and the symbol for kilowatts is kW.

WORD WISE

A KILOWATT IS EQUAL TO 1,000 WATTS.

Important Figures in Electricity

There are many important people in history who have contributed to our understanding of electricity. The ancient Greeks were the first to study electric forces. Thales was a Greek philosopher who lived about 2,500 years ago. He discovered static electricity by rubbing silk against a stone called amber. The buildup of electrons caused by the rubbing attracted feathers and bits of straw to the amber.

In the American colonies during the 1700s, Benjamin Franklin studied electricity. His experiments led to the invention of the lightning rod. That metal rod is used to protect buildings from lightning.

Benjamin Franklin invented many things in his lifetime, including the lightning rod.

Thomas Edison invented the lightbulb, along with the phonograph to record sound and an electric pen to make copies.

In 1878, the American inventor Thomas Edison introduced the world to the lightbulb. For centuries, people relied on fire and eventually gas lamps to light up their homes. These could be very dangerous. Homes could catch fire and be destroyed quickly. The lightbulb and our ability to control electrical energy may have saved millions of lives to this day. Many more inventors developed new technologies based on Thales, Franklin, and Edison's work!

compareandcontrast

In what ways are lightbulbs and lightning rods similar? How are the two inventions different?

Glossary

atmosphere The air that surrounds a planet.

chemical A substance that is formed when two or more other substances act upon one another or that is used to produce a change in another substance.

device A piece of equipment used to serve a special purpose.

electron A very small elementary particle within an atom that has a negative electric charge and travels around the nucleus of an atom.

environment All the physical surroundings on Earth, including everything living and nonliving.

fossil fuel A fuel, such as coal, oil, or natural gas, formed in the earth from plant or animal remains.

neutron An uncharged atomic particle that is present in all known atomic nuclei except the hydrogen nucleus.

nucleus The central part of an atom.

philosopher A person who seeks wisdom.

power plant A building in which electric power is generated.

proton A tiny particle that occurs in the nucleus of every atom and has a positive electric charge.

solar power The energy from sunlight that is used to run machinery.

turbine An engine with a central driving shaft that spins around. The shaft has winglike parts that are whirled around by the pressure of water, steam, or gas.

waterpower The energy from moving water that is used to run machinery.

windmill A mill or a machine worked by the wind turning sails or vanes at the top of a tower.

For More Information

Books

Bailey, Jacqui. *Let's Investigate Electricity.* St. Catharines, Ontario, Canada: Crabtree Publishing Company, 2021.

Gardner, Jane P. *Electricity*. Minneapolis, MN: Bearport Publishing, 2023.

Williams, Victoria. *How It Works: Electricity.* London, UK: Usborne Publishing Limited, 2023.

Websites

Britannica Kids: Electricity
www.kids.britannica.com/kids/article/electricity/353091
Learn more about electricity and watch fun videos on the Britannica Kids website.

Energy Kids: Science of Electricity
www.eia.gov/kids/energy-sources/electricity/science-of-electricity.php
View diagrams and read more information about electricity on the official U.S. Energy Information Administration website for kids.

Publisher's note to educators and parents: Our editors have carefully reviewed these websites to ensure that they are suitable for students. Many websites change frequently, however, and we cannot guarantee that a site's future contents will continue to meet our high standards of quality and educational value. Be advised that students should be closely supervised whenever they access the internet.

Index